DeltaScience ContentReaders

Weathering and Erosion

Contents

Preview the Book 2
What Are Landforms? 3
   Earth's Landforms 4
      Mountains 4
      Valleys 4
      Plains 5
      Plateaus 5

Compare and Contrast 6
What Are Weathering and Erosion? 7
   Weathering 8
   Erosion 10

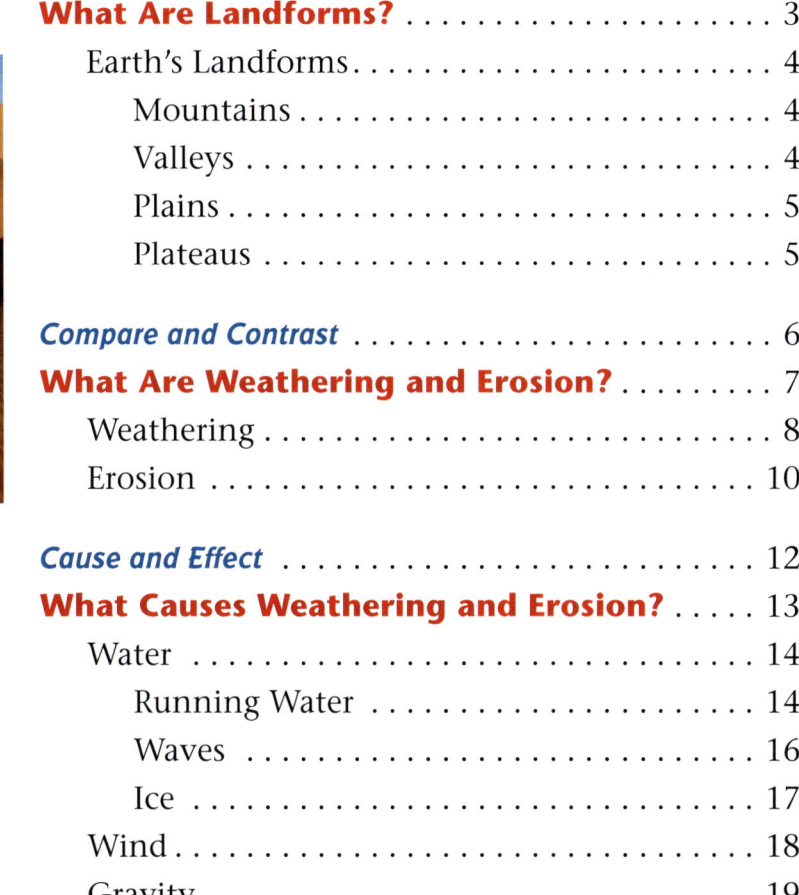

Cause and Effect 12
What Causes Weathering and Erosion? 13
   Water 14
      Running Water 14
      Waves 16
      Ice 17
   Wind 18
   Gravity 19

Main Idea and Details 20
How Do People Change the Land? 21
   People and Land 22

Glossary 24

Build Reading Skills
Preview the Book

You read nonfiction books like this one to learn about new ideas. Be sure to look through, or *preview*, the book before you start to read.

First, look at the title, front cover, and table of contents. What do you guess you will read about? Think about what you already know about weathering and erosion.

Next, look through the book page by page. Read the headings and the words in bold type. Look at the pictures and captions. Notice that each new part of the book starts with a big photograph. What other special features do you find in the book?

Headings, captions, and other features of nonfiction books are like road signs. They can help you find your way through new information. Now you are ready to read!

What Are Landforms?

MAKE A CONNECTION
The land has many natural shapes, or features. Some places have hills or mountains. What is the land like where you live?

FIND OUT ABOUT
- four main kinds of landforms

VOCABULARY

landform, p. 4
topography, p. 4
mountain, p. 4
valley, p. 4
plain, p. 5
plateau, p. 5

Earth's Landforms

Landforms are natural shapes, or features, on Earth's surface. Together, all the landforms in an area are called the area's **topography**.

Mountains

A **mountain** is a landform that is much higher than the land around it. Some mountains have pointed tops, while others are more rounded. A group of mountains is called a mountain range. The Rocky Mountains are a mountain range in North America. A *hill* is like a mountain, but it is not as high.

Valleys

A **valley** is a landform that lies between hills or mountains. It is much lower than the land around it. Some valleys are V-shaped. Others are more U-shaped. The sides of a valley are called walls. The bottom of a valley is called the floor. *Canyons* are deep valleys with steep walls.

▲ A mountain is a landform that is much higher than the land around it.

▲ A valley is a low landform between hills or mountains.

4

▲ A plain is a landform that is wide and flat.

▲ A plateau is a flat landform that is higher than the land around it.

Plains

A **plain** is a landform that is wide and flat. Plains are areas of land with very few hills or valleys. The Great Plains covers much of the central United States and parts of Canada and Mexico. Other plains are much smaller. A *floodplain* is a plain that lies along the sides of a river.

Plateaus

A **plateau** is a landform that is flat like a plain. But a plateau is higher than the land around it. Some plateaus are very large. For example, the Colorado Plateau stretches across much of Colorado, Utah, New Mexico, and Arizona. A *mesa* is a small plateau. A small mesa with very steep sides is called a *butte*.

 What is a landform? Name four kinds.

REFLECT ON READING
You previewed pictures, captions, and other book features before reading. Tell how one picture and caption helped you better understand a landform.

APPLY SCIENCE CONCEPTS
Find an example of a landform you have just read about. You can look in books or on the Internet. Talk to a partner about the name of the landform, its location, and its main features.

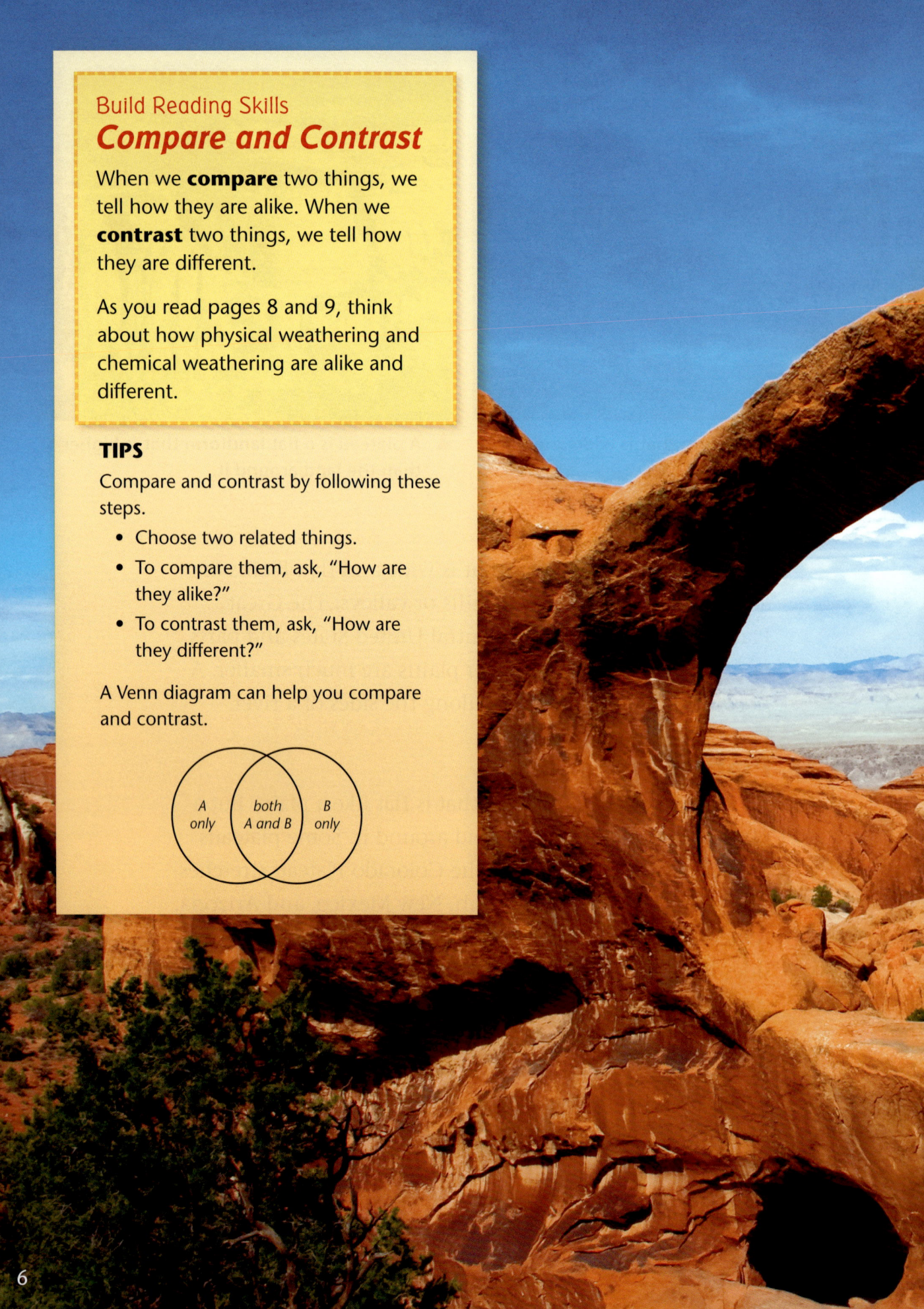

Build Reading Skills
Compare and Contrast

When we **compare** two things, we tell how they are alike. When we **contrast** two things, we tell how they are different.

As you read pages 8 and 9, think about how physical weathering and chemical weathering are alike and different.

TIPS

Compare and contrast by following these steps.

- Choose two related things.
- To compare them, ask, "How are they alike?"
- To contrast them, ask, "How are they different?"

A Venn diagram can help you compare and contrast.

What Are Weathering and Erosion?

MAKE A CONNECTION
Arches National Park in Utah has many natural sandstone arches. How do you think they formed?

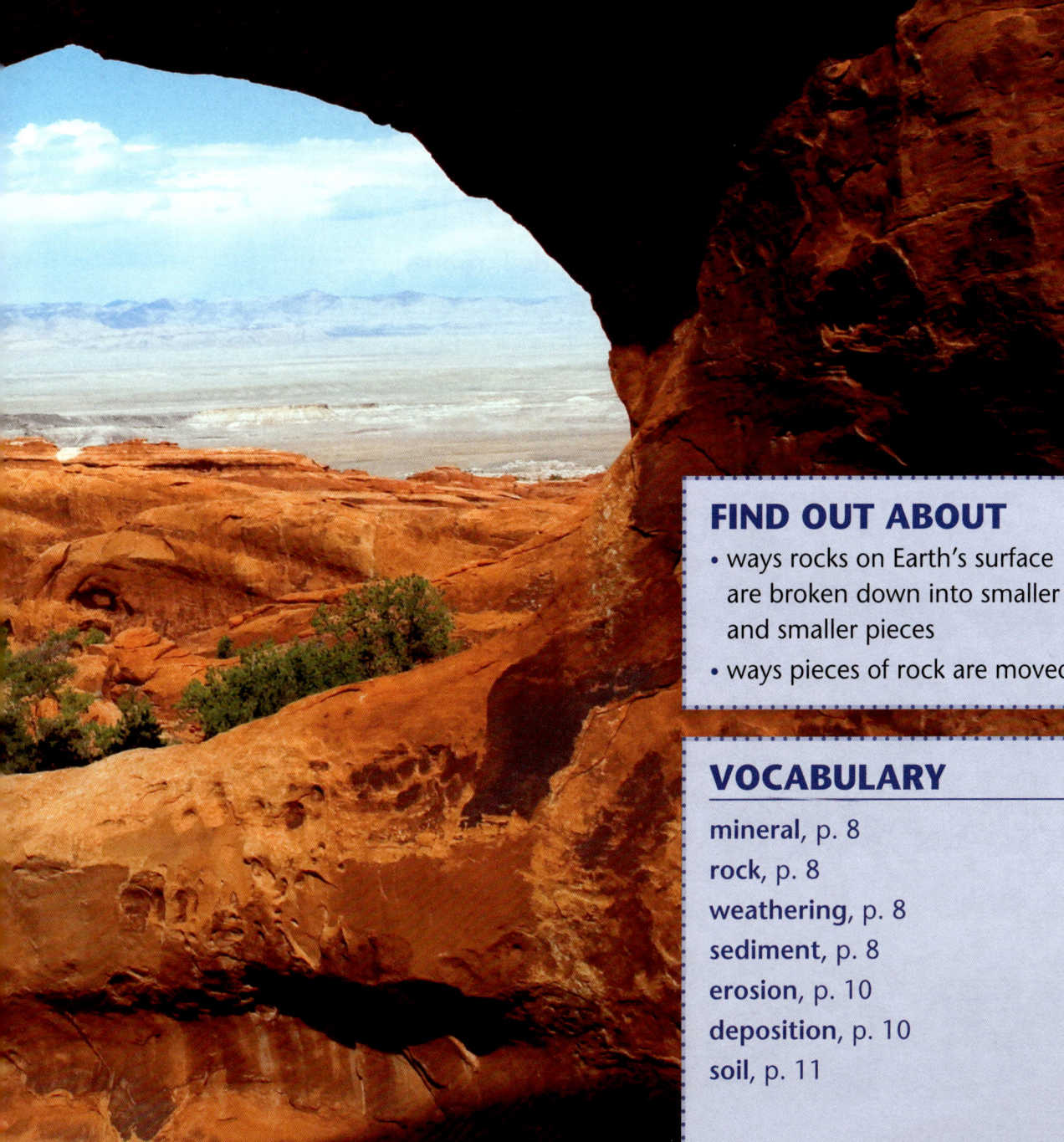

FIND OUT ABOUT
- ways rocks on Earth's surface are broken down into smaller and smaller pieces
- ways pieces of rock are moved

VOCABULARY

mineral, p. 8
rock, p. 8
weathering, p. 8
sediment, p. 8
erosion, p. 10
deposition, p. 10
soil, p. 11

7

Natural processes change the landforms on Earth's surface. Some changes may take millions of years. Other changes may take only a few minutes. Some changes build up landforms. Other changes break down landforms.

Weathering

Landforms are made up of materials such as soil, minerals, and rocks. **Minerals** are solid, nonliving materials found in nature. **Rocks** are made of minerals.

Weathering is the breaking down of minerals and rocks into smaller pieces. The smaller pieces are called **sediment**. Water, wind, temperature changes, and living things can cause weathering. You will read more about causes of weathering on pages 14–19.

Weathering can be physical or chemical. Physical weathering changes the size or shape of rocks. Rocks get smaller as they are scraped, pounded, rubbed, or split. For example, tree roots can grow into small cracks in a rock. As the roots grow, they split the rock apart.

This is a close-up photo of sand. Sand is an example of sediment. ▼

Growing tree roots can cause physical weathering. They can break rocks into smaller pieces. ▼

◀ The orange rust on these rocks formed because of chemical weathering.

Chemical weathering changes what rocks are made of. It can change minerals in rocks into different minerals. This can make rocks crumble into sediment.

Most chemical weathering happens when rocks are in contact with water or moist air. Materials in the water or air can join with materials in the rocks and change them. For example, suppose a stream flows over some rocks. Oxygen that is in the water may join with iron in those rocks. A reddish brown material called rust forms. Rust can make rocks soft and crumbly.

Living things can cause chemical weathering, too. Lichens are living things that often grow on rocks. Weak acids made by lichens can slowly break rocks down.

 **What is weathering?**

Erosion

Weathered rock, or sediment, often gets moved. **Erosion** is the movement of sediment. Erosion can happen very quickly. Or it can take thousands of years. Erosion breaks down landforms.

A force, such as a push or a pull, is needed to erode sediment. The force might come from moving water or from blowing wind, for example. More force is needed to erode larger pieces of sediment. You will read more about causes of erosion on pages 14–19.

Deposition is the dropping of eroded sediment in a new place. Suppose sediment is being carried by moving water or wind. The sediment gets deposited when the water or wind slows down. Larger pieces of sediment are dropped first. Smaller pieces of sediment are carried farther before they are dropped. Deposition builds up landforms.

▲ Strong wind can pick up and move loose sand. This is an example of erosion.

▲ Sediment was dropped when the river slowed. This is an example of deposition.

◀ Soil is made of sediment, water, air, and humus.

Deposited sediment may become part of soil. **Soil** is a mixture of sediment, water, air, and humus. Humus is the broken-down wastes and remains of living things.

The kinds of rocks in a place affect its soil. The sediment in soil comes from weathered rock. And different rocks have different minerals. So different soils have different kinds and amounts of minerals.

The average weather, or climate, in a place affects its soil, too. Soils in humid climates often have more humus than soils in dry climates.

The topography of a place also affects its soil. Steep areas often have less soil than flatter areas.

 What is erosion? Give an example.

REFLECT ON READING
Make a Venn diagram like the one on page 6. Use the Venn diagram to show how physical weathering and chemical weathering are alike and different.

APPLY SCIENCE CONCEPTS
Imagine that a waterfall flows over rocks for many years. The water slowly breaks down the rocks. Do you think this is weathering, erosion, or both? Write about your ideas in your science notebook.

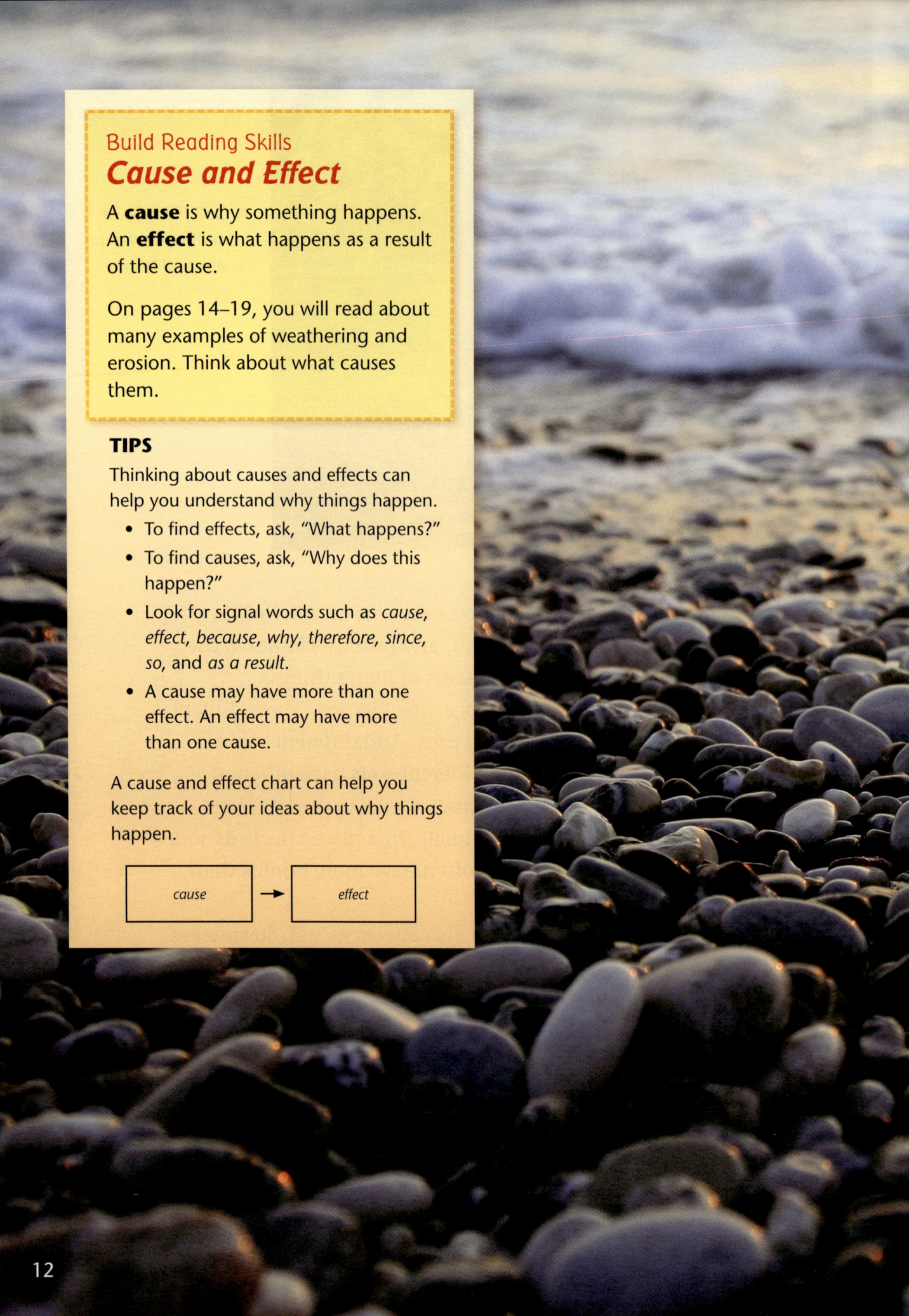

Build Reading Skills
Cause and Effect

A **cause** is why something happens. An **effect** is what happens as a result of the cause.

On pages 14–19, you will read about many examples of weathering and erosion. Think about what causes them.

TIPS

Thinking about causes and effects can help you understand why things happen.
- To find effects, ask, "What happens?"
- To find causes, ask, "Why does this happen?"
- Look for signal words such as *cause, effect, because, why, therefore, since, so,* and *as a result*.
- A cause may have more than one effect. An effect may have more than one cause.

A cause and effect chart can help you keep track of your ideas about why things happen.

| cause | → | effect |

What Causes Weathering and Erosion?

MAKE A CONNECTION
These beach stones are rounded and smooth. But they were not always this way. What do you think changed the shape of the stones?

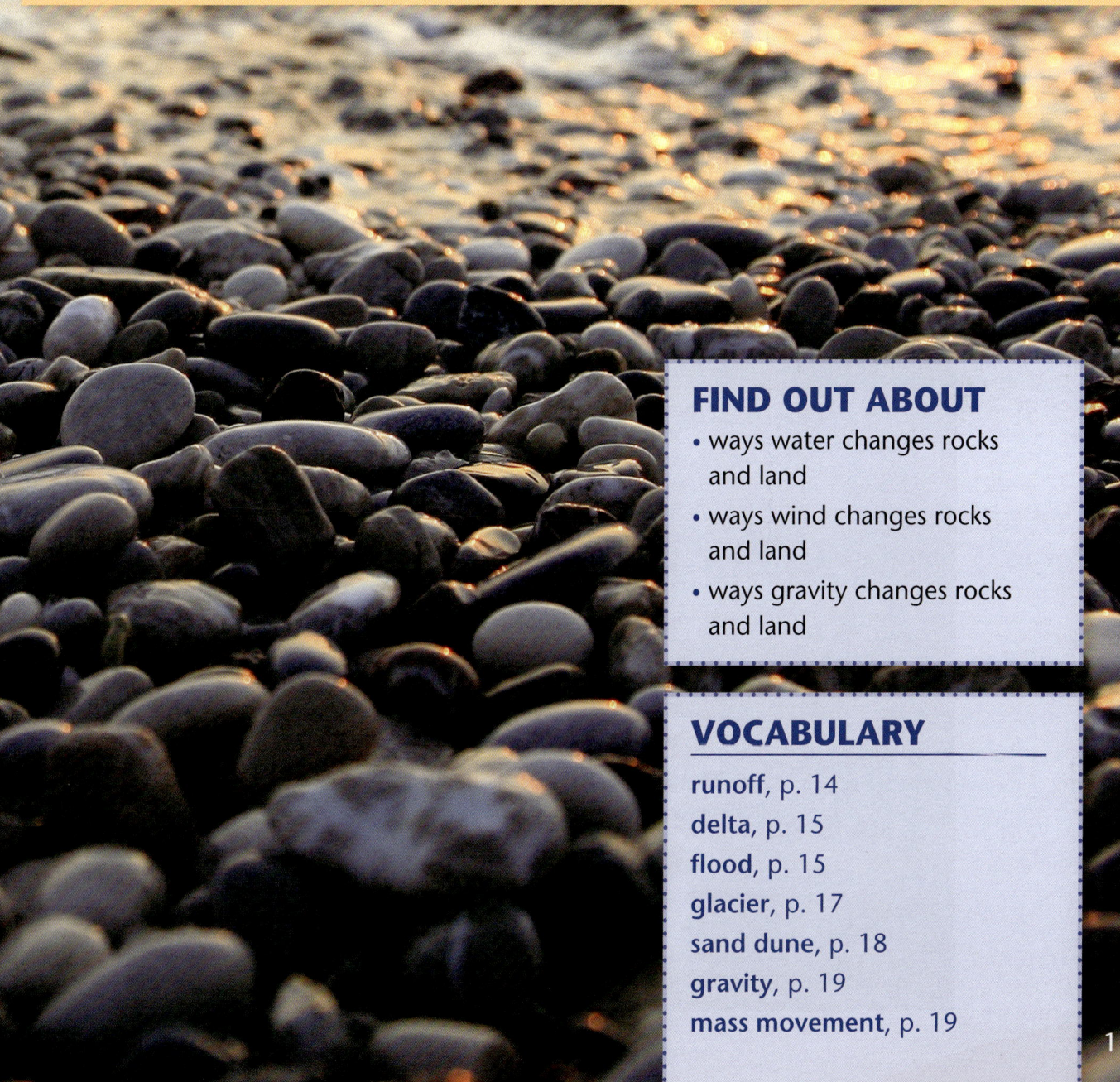

FIND OUT ABOUT
- ways water changes rocks and land
- ways wind changes rocks and land
- ways gravity changes rocks and land

VOCABULARY
runoff, p. 14
delta, p. 15
flood, p. 15
glacier, p. 17
sand dune, p. 18
gravity, p. 19
mass movement, p. 19

Water

Water causes most of the weathering, erosion, and deposition on Earth.

Running Water

Water that flows downhill is called running water. Examples of running water are rivers, streams, and runoff. **Runoff** is rain or melted snow that does not sink into the ground. Runoff flows downhill over the land into small streams. The streams flow into rivers. The end of a river is called its *mouth*. At its mouth, a river flows into a larger body of water such as an ocean.

Running water weathers rock and erodes sediment from the land under it. The steeper the land, the faster the water moves and the more erosion it causes. More water in a river also causes more erosion. Over time, rivers can carve valleys and canyons out of rock and soil.

Runoff, streams, and rivers erode sediment from the land. ▼

▲ The Mississippi River deposits sediment where it meets the ocean. This picture of the Mississippi delta was taken from space.

A river deposits sediment wherever it slows down, such as at its mouth. The sediment deposited at a river's mouth may make a landform called a **delta**. From above, a delta may look like a triangle or a bird's foot.

Rivers also deposit sediment when they overflow and cause a flood. A **flood** is water covering land that is usually dry. Runoff from melting snow or heavy rain can lead to flooding. Rivers often flood during the same months each year. The sediment deposited by a flood has minerals that plants need. So the land on a floodplain is good for farming.

A river may have bends or curves called meanders. The river once followed a straighter path. But it eroded sediment from its sides, or banks, in some places. And it deposited sediment in other places. This is why the river's path changed.

15

Waves

A shoreline is a place where land meets water. Waves at a shoreline can destroy land. They also can build land up.

Pounding waves can weather and erode a shoreline. Steep walls called cliffs may form. Waves also weather smaller rocks along the shoreline. The moving water grinds rocks against one another. Pieces of rock break off. Over time, rocks can become smooth, rounded pebbles or sand. Waves carry sand and other sediment away from some parts of the shoreline.

Waves also deposit sediment. A beach is a deposit of sediment along a shoreline. Ocean waves sometimes deposit sediment a short distance from shore. Ridges called sandbars form. If a sandbar gets large enough, it can become a landform called a *barrier island*.

▲ Ocean waves can weather and erode the rocks at a shoreline.

▲ A barrier island can form where ocean waves deposit sediment a short distance from shore.

▲ A glacier is a large body of ice that slowly moves over the land.

Ice

Frozen water, or ice, also causes changes to Earth's rocks and land. For example, water gets into cracks in rocks and then freezes. When water freezes, it takes up more space. This causes weathering. The ice widens the cracks in the rocks. Over time, the rocks split.

Glaciers are another example of how ice changes rocks and land. A **glacier** is a large body of ice that mostly stays frozen from year to year. Glaciers flow downhill over the land very slowly. Glaciers once covered more of Earth's surface than they do today.

A glacier causes erosion. It pushes soil and rocks ahead of it. Also, rocks freeze into the glacier and move with it.

A glacier causes weathering, too. Weathering happens when rocks and ice in the glacier grind against rocks on the land.

A glacier deposits sediment where parts of the glacier melt. A *moraine* is a long hill of sediment deposited by a glacier.

✅ Water can destroy landforms. Water also can build up landforms. Give an example of each.

Wind

Wind changes landforms through weathering, erosion, and deposition.

Wind causes weathering when it blows sand against rocks. The sand breaks tiny pieces off the rocks.

Wind causes erosion when it picks up and moves sand and smaller sediment. Larger pebbles and rocks are not usually moved by the wind. They are left behind.

When wind deposits sand, hills called **sand dunes** can form. Wind slows down when it hits an object such as a large rock or a clump of plants. Sand carried by the wind drops to the ground. Over time, the sand can build up and form a sand dune. The shape of a sand dune keeps changing. Wind deposits sand in some places and erodes sand from others.

✓ How does a sand dune form?

Sand dunes can form when wind deposits sand. ▶

◀ A rockfall is a kind of mass movement.

Gravity

Gravity is the force that makes things fall. It also is the force behind many forms of erosion and deposition. Gravity causes water to flow downhill. Gravity causes sediment to drop when moving water or wind slows.

Gravity also causes pieces of rock to fall and chunks of land to slide downhill. The movement of rocks and land by gravity is called **mass movement**. Here are three examples.

- *rockfall*   Pieces of loosened rock suddenly fall from a cliff or a steep slope.
- *landslide*   A whole chunk of land suddenly slides downhill.
- *creep*   The soil and rock on a slope move downhill very, very slowly.

✓ Gravity plays a role in many forms of erosion and deposition. Give two examples.

REFLECT ON READING

Make a cause and effect chart like the one on page 12. Write either "weathering" or "erosion" in the effect box. List some causes in the cause box.

APPLY SCIENCE CONCEPTS

Find pictures that show weathering and erosion. Look in magazines or on the Internet. With your class, make a poster that uses the pictures and tells about what they show, including the causes.

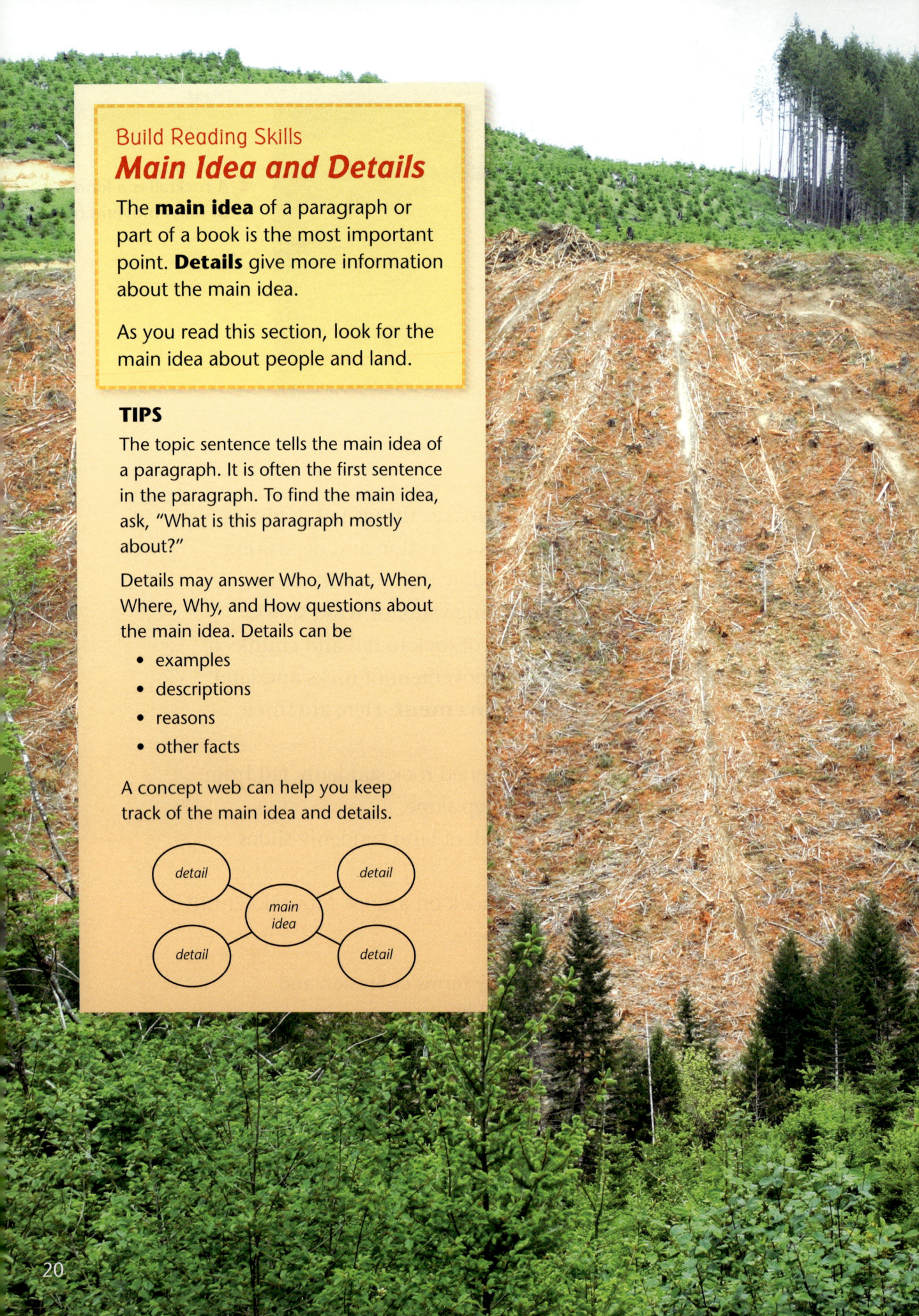

Build Reading Skills
Main Idea and Details

The **main idea** of a paragraph or part of a book is the most important point. **Details** give more information about the main idea.

As you read this section, look for the main idea about people and land.

TIPS

The topic sentence tells the main idea of a paragraph. It is often the first sentence in the paragraph. To find the main idea, ask, "What is this paragraph mostly about?"

Details may answer Who, What, When, Where, Why, and How questions about the main idea. Details can be

- examples
- descriptions
- reasons
- other facts

A concept web can help you keep track of the main idea and details.

How Do People Change the Land?

MAKE A CONNECTION

People cut down trees to get wood. Sometimes they cut down all the trees in an area. Why might people plant new trees here?

FIND OUT ABOUT
- some ways that people change the land
- how people are returning some land to its natural state

People and Land

People make changes to Earth's land. These changes can speed up or slow down natural processes such as weathering, erosion, and deposition.

People have built dams across some rivers. Dams can help control flooding downstream. Some dams are used to make electricity. But a dam changes the natural flow of water. This changes the amount of sediment carried by a river.

People sometimes remove plants from the land during building projects. But the roots of plants help hold soil in place. With fewer plants, more soil may get eroded by runoff and wind.

▲ A dam can be useful to people. But a dam changes how much sediment is moved by a river.

▲ The roots of plants help hold soil in place. Removing plants from the land speeds up erosion.

▲ This row of trees is a windbreak. It helps keep wind from eroding the soil. It slows the wind and changes its path.

People sometimes help slow the erosion of soil. For example, farmers may leave the stalks of plants in a field after the crop is cut. The roots of the plants help keep the soil in place. A farmer also can plant a row of trees near the field. This is called a windbreak. It helps stop the wind from blowing soil away.

Wetlands are low areas of land that are naturally covered with water. In the past, people sometimes filled in wetlands. This made dry land for building roads or houses. But wetlands can help slow erosion. They take in floodwater and block ocean waves. So people are working to return some wetlands to their natural state.

 Tell about something people do that can speed up erosion. Then tell about something people do that can slow erosion.

REFLECT ON READING

Make a concept web like the one on page 20. Use the web to keep track of ideas about the section you just read. Put the main idea in the middle. Then add details, such as examples.

APPLY SCIENCE CONCEPTS

How has the land near your home been changed by people? Write about one change in your science notebook. How do you think the change affected weathering or erosion there?

23

Glossary

delta (DEL-tuh) land at the mouth of a river formed from deposited sediment **(15)**

deposition (dep-uh-ZISH-uhn) the dropping of eroded sediment in a new place **(10)**

erosion (i-ROH-zhuhn) the movement of sediment **(10)**

flood (FLUHD) a large amount of water covering land that is usually dry **(15)**

glacier (GLAY-shur) a large body of ice that slowly moves downhill over land **(17)**

gravity (GRAV-uh-tee) the force that makes things fall to the ground **(19)**

landform (LAND-form) a natural shape, or feature, on Earth's surface **(4)**

mass movement (MAS MOOV-muhnt) the movement of rocks and land by gravity **(19)**

mineral (MIN-uhr-uhl) a solid, nonliving material found in nature; minerals make up rocks **(8)**

mountain (MOUN-tuhn) a landform that is much higher than the land around it **(4)**

plain (PLAYN) a landform that is wide and flat **(5)**

plateau (pla-TOH) a landform that is flat like a plain but is higher than the land around it **(5)**

rock (ROK) a natural solid that is made of one or more minerals **(8)**

runoff (RUHN-awf) rain or melted snow that does not sink into the ground but flows downhill across the land **(14)**

sand dune (SAND DOON) a hill of sand deposited and shaped by wind **(18)**

sediment (SED-uh-muhnt) pieces of minerals and rocks that are made by weathering **(8)**

soil (SOIL) a mixture of sediment, water, air, and once-living material called humus **(11)**

topography (tuh-POG-ruh-fee) all the landforms in an area **(4)**

valley (VAL-ee) a landform that lies between hills or mountains and is much lower than the land around it **(4)**

weathering (WE*TH*-uhr-ing) the breaking down of minerals and rocks into smaller pieces by water, wind, temperature changes, and living things **(8)**